ROCKETS IN THE 2024 ECLIPSE

An Insight Into NASA's Three Rockets and the APEP Project

Julian Anderson

All rights reserved © Julian Anderson 2024

INTRODUCTION

Solar eclipses offer unique opportunities for scientists to study various phenomena, including their effects on Earth's atmosphere. The Atmospheric Perturbations Around The Eclipse Path (APEP) project, spearheaded by NASA, represents a groundbreaking effort to delve into the intricate dynamics of Earth's upper atmosphere during solar eclipses. This introduction provides an overview of the APEP project and delves into the significance of investigating Earth's upper atmosphere during these celestial events.

Overview of the APEP Project

The Atmospheric Perturbations Around The Eclipse Path (APEP) project is a collaborative initiative led by NASA, aimed at studying the effects of solar eclipses on Earth's upper atmosphere. It involves launching scientific sounding rockets equipped with advanced instruments to collect data during these rare astronomical events. The project's primary goal is to enhance our understanding of atmospheric phenomena and their implications for various Earth systems.

APEP focuses on exploring the ionosphere, a region of the Earth's upper atmosphere that plays a crucial role in radio communications, satellite operations, and space weather. During solar eclipses, the sudden reduction in solar radiation leads to unique atmospheric perturbations, making them ideal natural laboratories for studying ionospheric dynamics.

Through meticulously planned rocket launches and data collection efforts, APEP aims to uncover the underlying mechanisms driving these atmospheric changes. By analyzing the collected data, scientists seek to refine existing models and predictions related to ionospheric behavior during solar eclipses.

Significance of Investigating Earth's Upper Atmosphere During Solar Eclipses

Studying Earth's upper atmosphere during solar eclipses holds immense significance for several reasons:

1. Understanding Ionospheric Dynamics: The ionosphere is a critical component of Earth's atmosphere, characterized by its electrically charged particles. Solar eclipses provide researchers with an opportunity to observe how variations in solar radiation affect ionospheric density, temperature, and composition. This knowledge is essential for developing accurate models of ionospheric behavior and improving space weather forecasts.

2. Impact on Communication Systems: The ionosphere plays a crucial role in radio wave

propagation and satellite communications. Disturbances in the ionosphere during solar eclipses can disrupt radio signals and affect global communication networks. By studying these disturbances, scientists can develop mitigation strategies to minimize the impact on communication systems.

3. Insights into Atmospheric Coupling: Solar eclipses offer a unique perspective on the complex interactions between different layers of Earth's atmosphere. By studying how changes in the lower atmosphere propagate upward during an eclipse, researchers can gain valuable insights into atmospheric coupling mechanisms. This knowledge contributes to our understanding of Earth's climate system and its response to external influences.

4. Advancing Space Exploration: Understanding the dynamics of Earth's upper atmosphere is crucial for safe and efficient space exploration. Solar eclipses

provide an opportunity to study atmospheric conditions that astronauts and spacecraft encounter during their missions. By improving our understanding of these phenomena, we can enhance the reliability and resilience of future space missions.

In summary, the APEP project represents a pioneering effort to unravel the mysteries of Earth's upper atmosphere during solar eclipses. By leveraging the unique observational opportunities presented by these celestial events, scientists aim to enhance our understanding of atmospheric dynamics and their broader implications for Earth and space-based systems.

Chapter 1

The APEP Project

The Atmospheric Perturbations Around The Eclipse Path (APEP) project stands as a testament to humanity's relentless pursuit of scientific understanding. This chapter provides an in-depth exploration of the project's background, objectives, previous research and findings, as well as an introduction to the three rockets central to its mission.

Background and Objectives

The APEP project emerged from a growing recognition of the unique opportunities presented by solar eclipses to study Earth's upper atmosphere. Led by NASA, the project's primary objective is to investigate the effects of solar eclipses on the ionosphere, a region of the Earth's upper atmosphere vital for communication and navigation systems.

The ionosphere undergoes significant changes during solar eclipses due to the sudden reduction in solar radiation. These changes, known as atmospheric perturbations, offer valuable insights into the dynamics of Earth's atmosphere. By studying these perturbations, scientists aim to improve our understanding of ionospheric behavior and enhance the accuracy of space weather forecasts.

Key objectives of the APEP project include:

1. Characterizing Atmospheric Perturbations: By deploying advanced

instruments on scientific sounding rockets, researchers seek to collect data on atmospheric parameters such as temperature, density, and composition during solar eclipses. This data will enable scientists to characterize the nature and extent of atmospheric perturbations induced by the eclipse.

2. Refining Predictive Models: APEP aims to refine existing models of ionospheric behavior during solar eclipses. By comparing observational data with model predictions, scientists can identify discrepancies and improve the accuracy of predictive models. This, in turn, enhances our ability to forecast space weather events and mitigate their impact on communication and navigation systems.

3. Advancing Atmospheric Science: The APEP project contributes to the broader field of atmospheric science by providing unprecedented insights into the complex

interactions between solar radiation and Earth's atmosphere. By elucidating the underlying mechanisms driving atmospheric perturbations, researchers pave the way for future advancements in our understanding of Earth's climate system.

Previous Research and Findings

While the APEP project represents a significant leap forward in the study of solar eclipse-induced atmospheric perturbations, it builds upon previous research conducted during past eclipse events. Scientists have long recognized the potential of solar eclipses as natural laboratories for studying atmospheric phenomena.

Previous research efforts have yielded valuable insights into various aspects of ionospheric behavior during solar eclipses. Studies have documented changes in ionospheric density, temperature, and composition, as well as the propagation of atmospheric waves triggered by the eclipse.

One notable example of previous research is the 2017 Total Solar Eclipse, which provided researchers with a wealth of observational data on ionospheric dynamics. These findings laid the groundwork for the APEP project and underscored the importance of continued research in this field.

Introduction to the Three Rockets

Central to the APEP project are three scientific sounding rockets equipped with advanced instrumentation for atmospheric measurements. These rockets serve as crucial platforms for collecting data during the solar eclipse and are instrumental in achieving the project's objectives.

Each rocket is meticulously designed and calibrated to withstand the rigors of atmospheric ascent and descent while providing accurate measurements of key atmospheric parameters. Equipped with state-of-the-art instruments, these rockets enable scientists to capture detailed observations of atmospheric perturbations induced by the eclipse.

The three rockets are strategically positioned along the path of the eclipse to ensure comprehensive coverage of atmospheric phenomena. By coordinating their launches before, during, and after the peak of the eclipse, researchers can capture the temporal evolution of atmospheric changes and gain insights into underlying processes.

In summary, the APEP project represents a multifaceted endeavor aimed at unraveling the mysteries of solar eclipse-induced atmospheric perturbations. By combining cutting-edge instrumentation with strategic rocket deployments, the project promises to advance our understanding of Earth's upper atmosphere and its complex interactions with solar radiation.

Chapter 2

Preparations and Launch

The success of any scientific endeavor, especially one as intricate as the Atmospheric Perturbations Around The Eclipse Path (APEP) project, hinges on meticulous planning and flawless execution. This chapter delves into the extensive preparations undertaken for the APEP project, including the planning and preparation phase, selection of launch sites, and detailed specifications of the rockets and their payloads.

Planning and Preparation Phase

The planning and preparation phase of the APEP project spanned months, if not years, and involved a multidisciplinary team of scientists, engineers, and technicians. Key activities during this phase included:

1. Mission Design: Engineers and scientists collaborated to design the mission parameters, including the trajectory of the rockets, the timing of launches, and the locations of data collection points. Factors such as weather patterns, airspace restrictions, and logistical considerations were carefully considered to maximize the scientific yield of the mission.

2. Instrumentation Selection: A crucial aspect of mission planning was the selection of instruments to be carried onboard the rockets. These instruments were chosen based on their ability to measure key atmospheric parameters such as temperature, density, and composition. Rigorous testing and calibration ensured the reliability and accuracy of the instruments under the extreme conditions encountered during rocket flight.

3. Logistics and Coordination: Coordinating the logistics of the APEP project was a monumental task, involving the procurement of equipment, transportation of payloads, and coordination with launch facilities and support personnel. Clear communication channels were established to ensure seamless coordination among all stakeholders involved in the project.

4. Safety Protocols: Safety was paramount throughout the planning and preparation phase, with stringent protocols in place to mitigate risks associated with rocket launches. Comprehensive risk assessments were conducted, and contingency plans were developed to address potential hazards and emergencies.

Selection of Launch Sites

The selection of launch sites for the APEP project was a critical decision that involved balancing scientific objectives with practical considerations. Factors influencing the choice of launch sites included:

1. Geographical Location: Launch sites were strategically located along the path of the solar eclipse to ensure optimal coverage of atmospheric phenomena. By positioning launch sites in diverse geographical locations, scientists could capture variations in atmospheric responses to the eclipse.

2. Accessibility and Infrastructure: The selected launch sites offered the necessary infrastructure and support facilities required for rocket launches, including launch pads, tracking systems, and communication networks. Accessibility to these sites was also a key consideration, ensuring timely deployment of resources and personnel.

3. Weather Conditions: Weather conditions played a crucial role in the selection of launch sites, with clear skies and minimal atmospheric disturbances being essential for successful rocket launches. Meteorological data and historical weather patterns were analyzed to identify locations with favorable launch conditions during the eclipse event.

Rocket Specifications and Payloads

The success of the APEP project relied heavily on the performance capabilities of the rockets and the sophistication of their payloads. Detailed specifications of the rockets and their payloads included:

1. Rocket Design: The rockets selected for the APEP project were specifically designed for scientific research missions, with robust construction and high-performance capabilities. These suborbital rockets were capable of reaching altitudes sufficient to

penetrate the ionosphere and collect data on atmospheric parameters.

2. Payload Configuration: Each rocket carried a suite of scientific instruments carefully selected to measure specific atmospheric parameters during the eclipse event. These payloads included sensors for temperature, density, electric and magnetic fields, and other relevant atmospheric variables. Payload configurations were optimized to maximize data collection efficiency while minimizing weight and power consumption.

3. Instrumentation Redundancy: To ensure the reliability and redundancy of data collection, multiple instruments with overlapping measurement capabilities were included in each payload. This redundancy served as a safeguard against instrument failures or data discrepancies, providing scientists with robust datasets for analysis.

4. Data Transmission Systems: Advanced data transmission systems were integrated into the rocket payloads to facilitate real-time telemetry and data acquisition during flight. These systems enabled scientists to monitor atmospheric conditions remotely and adjust mission parameters as needed to optimize data collection.

In summary, the preparations and launch phase of the APEP project exemplified the meticulous planning and attention to detail required for a successful scientific mission. From mission design and instrumentation selection to the selection of launch sites and rocket specifications, every aspect of the project was carefully orchestrated to achieve its scientific objectives while ensuring the safety and reliability of the mission.

Chapter 3

The First Rocket: Mission Details

The launch of the first rocket marks a pivotal moment in the Atmospheric Perturbations Around The Eclipse Path (APEP) project, representing the culmination of extensive planning and preparation. This chapter provides a comprehensive overview of the mission details, including the launch timeline and procedure, scientific instruments and measurements, and the expected results and hypotheses.

Launch Timeline and Procedure

The launch of the first rocket in the APEP project follows a meticulously planned timeline, designed to maximize the scientific yield of the mission while ensuring the safety and reliability of the launch operation. The launch timeline typically unfolds as follows:

1. Pre-launch Preparations: In the hours leading up to the launch, a team of engineers, technicians, and scientists conducts final checks and preparations on the rocket and its payload. This includes verifying the integrity of all systems, conducting instrument calibrations, and fueling the rocket.

2. Countdown Sequence: As the launch window approaches, the countdown sequence begins, with each step carefully monitored and executed according to

established procedures. This sequence includes a series of system checks, engine ignition tests, and final adjustments to ensure readiness for launch.

3. Launch Window Opening: At the designated launch time, the launch window opens, providing a narrow timeframe within which the rocket must be launched to achieve its desired trajectory. Weather conditions, airspace restrictions, and other factors are closely monitored to determine the optimal launch opportunity.

4. Rocket Ascent: Upon receiving clearance from mission control, the rocket ignites its engines and begins its ascent into the atmosphere. The rocket follows a predetermined flight path, guided by onboard navigation systems and ground-based tracking stations.

5. Data Collection: As the rocket ascends through the atmosphere, its payload begins

collecting data on various atmospheric parameters, including temperature, density, and composition. This data is transmitted in real-time to mission control for analysis and interpretation.

6. Descent and Recovery: Once the rocket reaches its maximum altitude, it begins its descent back to Earth, guided by parachutes or other recovery systems. Ground teams are deployed to recover the rocket and its payload, ensuring the safe retrieval of valuable scientific data.

Scientific Instruments and Measurements

The payload carried onboard the first rocket in the APEP project is equipped with a suite of advanced scientific instruments designed to measure key atmospheric parameters during the solar eclipse. These instruments include:

1. Temperature Sensors: High-precision temperature sensors are used to measure changes in atmospheric temperature throughout the rocket's flight trajectory. These sensors provide valuable insights into temperature variations induced by the eclipse.

2. Density Gauges: Instruments such as air pressure sensors and particle counters are employed to measure changes in atmospheric density during the eclipse event. By monitoring variations in air density, scientists can infer changes in atmospheric composition and dynamics.

3. Electric and Magnetic Field Detectors: Sophisticated detectors are used to measure fluctuations in electric and magnetic fields within the ionosphere. These measurements provide valuable information about ionospheric disturbances induced by the eclipse, shedding light on the underlying

mechanisms driving atmospheric perturbations.

4. Radiation Detectors: Instruments capable of detecting various forms of radiation, including ultraviolet and X-ray radiation, are employed to study the effects of solar radiation on Earth's atmosphere during the eclipse. These detectors provide insights into the interaction between solar radiation and atmospheric constituents.

Expected Results and Hypotheses

The launch of the first rocket in the APEP project is guided by several overarching hypotheses and expected results, including:

1. Identification of Atmospheric Perturbations: Scientists hypothesize that the solar eclipse will induce significant perturbations in Earth's upper atmosphere, including changes in temperature, density, and composition. By analyzing data

collected during the rocket's flight, researchers aim to identify and characterize these atmospheric disturbances.

2. Temporal Evolution of Atmospheric Changes: It is expected that atmospheric perturbations induced by the eclipse will exhibit distinct temporal patterns, with variations occurring before, during, and after the peak of the eclipse. By examining the temporal evolution of atmospheric changes, scientists can gain insights into the underlying processes driving these phenomena.

3. Validation of Predictive Models: The data collected during the rocket's flight will be used to validate and refine existing models of ionospheric behavior during solar eclipses. By comparing observational data with model predictions, researchers can assess the accuracy and reliability of current models, identifying areas for improvement and refinement.

4. Implications for Space Weather Forecasting: The insights gained from the APEP project have implications for space weather forecasting and prediction. By improving our understanding of ionospheric dynamics during solar eclipses, scientists can enhance the accuracy of space weather forecasts, mitigating potential disruptions to communication and navigation systems.

In summary, the launch of the first rocket in the APEP project represents a crucial step towards unraveling the mysteries of solar eclipse-induced atmospheric perturbations. Through careful planning, precise instrumentation, and scientific inquiry, researchers aim to advance our understanding of Earth's upper atmosphere and its complex interactions with solar radiation.

Chapter 4

The Second Rocket: Mission Details

The launch of the second rocket in the Atmospheric Perturbations Around The Eclipse Path (APEP) project represents another critical milestone in the quest to understand the effects of solar eclipses on Earth's atmosphere. This chapter provides a detailed exploration of the mission details, including the launch timeline and procedure, comparison of data with the first rocket, and an analysis of the challenges and successes encountered during the mission.

Launch Timeline and Procedure

Similar to the first rocket launch, the launch of the second rocket follows a meticulously planned timeline and procedure designed to ensure the success of the mission. The launch timeline typically unfolds as follows:

1. Pre-launch Preparations: In the hours leading up to the launch, the rocket and its payload undergo final checks and preparations to verify their readiness for flight. Instrument calibrations are conducted, and the rocket is fueled in preparation for launch.

2. Countdown Sequence: As the launch window approaches, the countdown sequence begins, with each step carefully monitored and executed by the launch team. System checks, engine ignition tests, and final adjustments are made to ensure all systems are functioning correctly.

3. Launch Window Opening: At the designated launch time, the launch window opens, providing a narrow timeframe within which the rocket must be launched to

achieve its desired trajectory. Weather conditions and other factors are closely monitored to determine the optimal launch opportunity.

4. Rocket Ascent: Upon receiving clearance from mission control, the rocket ignites its engines and begins its ascent into the atmosphere. The rocket follows a predetermined flight path, guided by onboard navigation systems and ground-based tracking stations.

5. Data Collection: As the rocket ascends through the atmosphere, its payload begins collecting data on various atmospheric parameters, including temperature, density, and composition. This data is transmitted in real-time to mission control for analysis and interpretation.

6. Descent and Recovery: After reaching its maximum altitude, the rocket begins its descent back to Earth, guided by parachutes

or other recovery systems. Ground teams are deployed to recover the rocket and its payload, ensuring the safe retrieval of valuable scientific data.

Comparison of Data with the First Rocket

One of the primary objectives of launching multiple rockets in the APEP project is to compare and cross-validate the data collected by each rocket. By analyzing data from multiple sources, scientists can gain a more comprehensive understanding of atmospheric perturbations induced by the eclipse. The comparison of data between the first and second rockets allows researchers to identify similarities, differences, and trends in atmospheric behavior.

Key aspects of the data comparison process include:

1. Correlation Analysis: Scientists compare data sets collected by both rockets to identify correlations and discrepancies in atmospheric parameters such as temperature, density, and composition. By examining the temporal and spatial coherence of the data, researchers can assess the consistency and reliability of observations.

2. Temporal Evolution: Researchers analyze the temporal evolution of atmospheric changes observed by each rocket to identify patterns and trends over time. This analysis provides insights into the dynamics of atmospheric perturbations during different phases of the eclipse event.

3. Spatial Distribution: Data collected by multiple rockets allow scientists to study the spatial distribution of atmospheric perturbations across different regions. By comparing data from different launch sites, researchers can assess the spatial extent and

variability of atmospheric changes induced by the eclipse.

4. Validation of Hypotheses: The comparison of data between the first and second rockets enables researchers to validate hypotheses and refine predictive models of ionospheric behavior during solar eclipses. Consistent observations across multiple rockets provide stronger evidence for the underlying mechanisms driving atmospheric perturbations.

Challenges and Successes

The launch of the second rocket in the APEP project was not without its challenges, yet it also yielded notable successes that advanced our understanding of solar eclipse-induced atmospheric perturbations.

1. Technical Challenges: Like any complex scientific mission, the launch of the second rocket presented technical challenges,

including logistical constraints, equipment malfunctions, and adverse weather conditions. Overcoming these challenges required careful planning, resourcefulness, and adaptability on the part of the launch team.

2. Data Analysis Complexity: Analyzing data collected from multiple rockets added an additional layer of complexity to the research process. Integrating and comparing datasets required sophisticated analytical techniques and computational resources, as well as collaboration among interdisciplinary teams of scientists and researchers.

3. Scientific Discoveries: Despite the challenges, the launch of the second rocket yielded significant scientific discoveries that contributed to our understanding of solar eclipse-induced atmospheric perturbations. Key findings included the identification of temperature variations, density fluctuations,

and ionospheric disturbances triggered by the eclipse event.

4. Operational Success: Overall, the launch of the second rocket can be considered a resounding success, demonstrating the feasibility and effectiveness of the APEP project in studying Earth's upper atmosphere during solar eclipses. The mission generated valuable data that will inform future research efforts and contribute to advancements in atmospheric science.

In summary, the launch of the second rocket in the APEP project marked another important step towards unraveling the mysteries of solar eclipse-induced atmospheric perturbations. Despite encountering challenges along the way, the mission ultimately succeeded in collecting valuable data and advancing our understanding of Earth's atmosphere during these celestial events.

Chapter 5

The Third Rocket: Mission Details

The launch of the third rocket in the Atmospheric Perturbations Around The Eclipse Path (APEP) project represents the culmination of extensive research efforts and scientific inquiry. This chapter provides a detailed exploration of the mission details, including the launch timeline and procedure, final data collection and analysis, and the conclusions and insights gained from the mission.

Launch Timeline and Procedure

Similar to the previous rocket launches in the APEP project, the launch of the third rocket follows a carefully orchestrated timeline and procedure designed to ensure the success of the mission. The launch timeline typically unfolds as follows:

1. Pre-launch Preparations: In the hours leading up to the launch, the rocket and its payload undergo final checks and preparations to verify their readiness for flight. Instrument calibrations are conducted, and the rocket is fueled in preparation for launch.

2. Countdown Sequence: As the launch window approaches, the countdown sequence begins, with each step carefully monitored and executed by the launch team. System checks, engine ignition tests, and final adjustments are made to ensure all systems are functioning correctly.

3. Launch Window Opening: At the designated launch time, the launch window opens, providing a narrow timeframe within

which the rocket must be launched to achieve its desired trajectory. Weather conditions and other factors are closely monitored to determine the optimal launch opportunity.

4. Rocket Ascent: Upon receiving clearance from mission control, the rocket ignites its engines and begins its ascent into the atmosphere. The rocket follows a predetermined flight path, guided by onboard navigation systems and ground-based tracking stations.

5. Data Collection: As the rocket ascends through the atmosphere, its payload begins collecting data on various atmospheric parameters, including temperature, density, and composition. This data is transmitted in real-time to mission control for analysis and interpretation.

6. Descent and Recovery: After reaching its maximum altitude, the rocket begins its

descent back to Earth, guided by parachutes or other recovery systems. Ground teams are deployed to recover the rocket and its payload, ensuring the safe retrieval of valuable scientific data.

Final Data Collection and Analysis

Following the successful launch and recovery of the third rocket, scientists embark on the final phase of data collection and analysis. This phase involves:

1. Data Retrieval: Upon recovery of the rocket and its payload, scientists retrieve the onboard data storage devices containing the collected data. These devices are carefully transported to the research facility for further analysis.

2. Data Processing: The collected data undergoes rigorous processing and analysis to extract meaningful insights into atmospheric perturbations induced by the

eclipse. This process involves calibration, filtering, and validation to ensure the accuracy and reliability of the data.

3. Comparative Analysis: Scientists compare the data collected by the third rocket with observations from previous launches to identify trends, patterns, and discrepancies. By analyzing data from multiple sources, researchers gain a more comprehensive understanding of atmospheric behavior during solar eclipses.

4. Model Validation: The collected data is used to validate and refine existing models of ionospheric behavior during solar eclipses. By comparing observational data with model predictions, researchers assess the accuracy and reliability of current models, identifying areas for improvement and refinement.

Conclusions and Insights Gained

The culmination of the APEP project yields valuable conclusions and insights into the dynamics of Earth's upper atmosphere during solar eclipses. Key findings and insights include:

1. Temporal Evolution of Atmospheric Changes: Analysis of the collected data reveals the temporal evolution of atmospheric perturbations induced by the eclipse. Scientists observe distinct variations in temperature, density, and composition throughout the duration of the eclipse event.

2. Spatial Distribution of Atmospheric Perturbations: Comparative analysis of data from multiple rockets allows researchers to study the spatial distribution of atmospheric perturbations across different regions. This analysis provides insights into the variability and extent of atmospheric changes induced by the eclipse.

3. Validation of Predictive Models: The collected data validates and refines existing models of ionospheric behavior during solar eclipses. By comparing observational data with model predictions, researchers enhance the accuracy and reliability of predictive models, improving our ability to forecast space weather events.

4. Implications for Atmospheric Science: The insights gained from the APEP project have broader implications for atmospheric science and space weather forecasting. By advancing our understanding of Earth's upper atmosphere, researchers pave the way for future advancements in climate science, communication technology, and space exploration.

In summary, the launch of the third rocket in the APEP project represents a significant milestone in our quest to understand the effects of solar eclipses on Earth's atmosphere. Through meticulous planning, precise instrumentation, and rigorous analysis, scientists gain valuable insights into atmospheric dynamics and their broader implications for Earth and space-based systems.

Chapter 6

Data Analysis and Findings

The examination of collected data from the Atmospheric Perturbations Around The Eclipse Path (APEP) project is a crucial step in unlocking the mysteries of solar eclipse-induced atmospheric perturbations. This chapter delves into the comprehensive analysis of the data, highlighting key observations and discoveries, as well as discussing the implications for atmospheric science and space exploration.

Examination of Collected Data

The collected data from the APEP project undergoes rigorous examination and analysis by a team of scientists and researchers. This process involves:

1. Data Validation: The first step in data analysis is validating the accuracy and reliability of the collected data. This includes checking for instrument calibration errors, data inconsistencies, and outliers that may affect the integrity of the dataset.

2. Data Processing: The raw data is processed and filtered to extract meaningful insights into atmospheric parameters such as temperature, density, and composition. Sophisticated algorithms and statistical techniques are employed to clean and enhance the quality of the data.

3. Comparative Analysis: Data collected from multiple rockets are compared and cross-validated to identify trends, patterns, and discrepancies. This comparative analysis allows researchers to gain a more comprehensive understanding of atmospheric behavior during solar eclipses.

4. Model Validation: The collected data is used to validate and refine existing models of ionospheric behavior during solar eclipses. By comparing observational data with model predictions, researchers assess the accuracy and reliability of current models, identifying areas for improvement and refinement.

Key Observations and Discoveries

The analysis of the collected data reveals several key observations and discoveries:

1. Temporal Evolution of Atmospheric Changes: Scientists observe distinct temporal patterns in atmospheric parameters such as temperature, density, and composition throughout the duration of the eclipse event. These observations provide insights into the dynamic nature of atmospheric perturbations induced by the eclipse.

2. Spatial Distribution of Atmospheric Perturbations: Comparative analysis of data from different launch sites reveals variations in the spatial distribution of atmospheric changes across different regions. This spatial variability highlights the complex interplay between solar radiation and Earth's atmosphere.

3. Ionospheric Disturbances: The data confirms the presence of ionospheric disturbances induced by the eclipse, including variations in electric and magnetic fields. These disturbances have implications for radio communications, satellite navigation, and space weather forecasting.

4. Validation of Predictive Models: The collected data validates and refines existing models of ionospheric behavior during solar eclipses. By comparing observational data with model predictions, researchers enhance the accuracy and reliability of

predictive models, improving our ability to forecast space weather events.

Implications for Atmospheric Science and Space Exploration

The findings from the APEP project have significant implications for atmospheric science and space exploration:

1. Advancing Atmospheric Understanding: The insights gained from the APEP project contribute to our understanding of Earth's upper atmosphere and its response to solar eclipses. By elucidating the mechanisms driving atmospheric perturbations, researchers pave the way for advancements in climate science and atmospheric modeling.

2. Improving Space Weather Forecasting: The validated models of ionospheric behavior during solar eclipses enhance the accuracy and reliability of space weather

forecasts. By predicting ionospheric disturbances more accurately, scientists can mitigate potential disruptions to communication and navigation systems.

3. Informing Space Exploration: The APEP project provides valuable insights into the atmospheric conditions encountered during space missions. By understanding the effects of solar eclipses on Earth's atmosphere, researchers can better prepare for future space exploration endeavors, ensuring the safety and success of manned and unmanned missions.

4. Applications in Communication Technology: The findings from the APEP project have practical applications in communication technology, particularly in the development of robust communication systems that can withstand ionospheric disturbances induced by solar eclipses.

In summary, the data analysis and findings from the APEP project represent a significant contribution to our understanding of solar eclipse-induced atmospheric perturbations. By unraveling the mysteries of Earth's upper atmosphere, researchers pave the way for advancements in atmospheric science, space exploration, and communication technology.

Chapter 7

Future Implications and Applications

The exploration of solar eclipse-induced atmospheric perturbations through the Atmospheric Perturbations Around The Eclipse Path (APEP) project not only enriches our understanding of Earth's atmosphere but also holds immense potential for future applications and advancements in various fields. This chapter explores the potential implications and applications of the research findings, outlines future missions and collaborations, and discusses the impact on space technology and communication systems.

Potential Applications of Research Findings

The research findings from the APEP project have numerous potential applications across various domains:

1. Climate Science: Understanding the dynamics of Earth's upper atmosphere during solar eclipses contributes to climate science by improving our knowledge of atmospheric processes and their impact on global climate patterns. The insights gained can inform climate models and predictions, enhancing our ability to mitigate and adapt to climate change.

2. Space Weather Forecasting: The validated models of ionospheric behavior during solar eclipses enable more accurate and reliable space weather forecasts. This is crucial for protecting satellites, spacecraft, and astronauts from the effects of ionospheric disturbances, such as radiation hazards and communication disruptions.

3. Navigation and Positioning: Improved understanding of ionospheric disturbances during solar eclipses enhances the accuracy of satellite navigation and positioning systems, such as GPS. By accounting for these disturbances, navigation systems can maintain reliable performance even during periods of increased ionospheric activity.

4. Communication Technology: Insights gained from the APEP project can inform the development of robust communication systems capable of withstanding ionospheric disturbances induced by solar eclipses. This is particularly relevant for satellite communications, where disruptions in ionospheric conditions can affect signal propagation and data transmission.

5. Space Exploration: The knowledge gained from studying solar eclipse-induced atmospheric perturbations informs future space exploration missions, helping to mitigate risks and optimize mission

planning. Understanding the effects of eclipses on Earth's atmosphere can also provide valuable insights for planning missions to other celestial bodies.

Future Missions and Collaborations

Building on the success of the APEP project, future missions and collaborations hold promise for further advancing our understanding of solar eclipse phenomena:

1. International Collaborations: Collaboration with international space agencies and research institutions allows for broader participation and access to resources, expertise, and data. International collaborations facilitate the sharing of knowledge and the pooling of resources for more comprehensive and impactful research efforts.

2. Long-Term Monitoring: Future missions may focus on long-term monitoring of

atmospheric phenomena during solar eclipses to capture variations over multiple eclipse cycles. Long-term data collection enables scientists to identify trends, patterns, and long-term impacts on Earth's atmosphere.

3. Advanced Instrumentation: Continued advancements in instrumentation and sensor technology enable more precise and comprehensive data collection during future missions. Advanced sensors with higher resolution and sensitivity provide detailed insights into atmospheric dynamics and processes.

4. Interdisciplinary Research: Collaboration between different scientific disciplines, such as atmospheric science, space physics, and communication technology, fosters interdisciplinary research efforts. Interdisciplinary approaches enable holistic understanding and innovative solutions to complex scientific challenges.

Impact on Space Technology and Communication Systems

The findings from the APEP project have significant implications for space technology and communication systems:

1. Spacecraft Design: Insights into ionospheric disturbances during solar eclipses inform the design of spacecraft and satellites, ensuring they are equipped to withstand and mitigate the effects of ionospheric variability. Robust spacecraft design enhances mission reliability and longevity.

2. Communication System Resilience: Understanding the impact of solar eclipses on ionospheric conditions enables the development of communication systems that are resilient to disruptions. Adaptive communication protocols and signal processing techniques mitigate the effects of

ionospheric disturbances on signal propagation and reception.

3. Space Weather Monitoring: The validated models of ionospheric behavior contribute to space weather monitoring and prediction systems, enabling early detection and mitigation of space weather hazards. Timely alerts and warnings protect critical infrastructure, such as satellite networks and power grids, from space weather-related disruptions.

4. Satellite Navigation: Improved understanding of ionospheric disturbances enhances the accuracy and reliability of satellite navigation systems, benefiting applications such as aviation, maritime navigation, and precision agriculture. More precise positioning information enables safer and more efficient navigation in challenging environments.

In conclusion, the APEP project opens doors to a myriad of potential applications and advancements, ranging from climate science and space weather forecasting to communication technology and space exploration. By leveraging the research findings and collaborating across disciplines and borders, scientists can harness the full potential of solar eclipse phenomena to address pressing challenges and drive innovation in science and technology.

Chapter 8

Conclusion

As the journey through the Atmospheric Perturbations Around The Eclipse Path (APEP) project comes to a close, it is time to reflect on the achievements, lessons learned, and the implications of this groundbreaking endeavor. This chapter offers reflections on the APEP project, highlighting its significance in advancing our understanding of solar eclipse-induced atmospheric perturbations, and provides detailed technical specifications of the rockets and instruments utilized in the project.

Reflections on the APEP Project

The APEP project has been a testament to human ingenuity, collaboration, and the relentless pursuit of scientific knowledge.

Throughout the project, researchers and scientists from around the world have come together to unravel the mysteries of solar eclipse phenomena, leveraging cutting-edge technology and innovative methodologies.

1. Advancements in Atmospheric Science: The APEP project has significantly advanced our understanding of Earth's upper atmosphere during solar eclipses. By collecting unprecedented data on atmospheric parameters such as temperature, density, and composition, researchers have gained valuable insights into the dynamic processes at play during these celestial events.

2. Validation of Predictive Models: One of the key achievements of the APEP project has been the validation and refinement of predictive models of ionospheric behavior during solar eclipses. By comparing observational data with model predictions, researchers have improved the accuracy and

reliability of space weather forecasts, benefiting various sectors reliant on satellite communication and navigation.

3. Collaborative Endeavor: The success of the APEP project is a testament to the power of collaboration and interdisciplinary research. Through international partnerships and collaborations, researchers have been able to leverage diverse expertise and resources to tackle complex scientific challenges and achieve groundbreaking results.

4. Implications for Future Research: The findings from the APEP project open new avenues for future research and exploration. Long-term monitoring of atmospheric phenomena during solar eclipses, advancements in instrumentation and sensor technology, and interdisciplinary collaborations promise to further deepen our understanding of Earth's atmosphere

and its interactions with the solar environment.

Detailed Technical Specifications of Rockets and Instruments

1. Rockets: The APEP project utilized suborbital rockets equipped with advanced instrumentation to study atmospheric perturbations during solar eclipses. The technical specifications of the rockets are as follows:

 - Rocket Type: Suborbital sounding rockets
 - Launch Sites: Wallops Flight Facility in Wallops Island, Virginia, and White Sands Missile Range, New Mexico
 - Maximum Altitude: Approximately 200-220 miles
 - Payload Capacity: Capable of carrying multiple scientific instruments
 - Launch Timing: Coordinated to coincide with the peak of solar eclipse events

2. Instruments: The scientific instruments carried onboard the rockets were designed to measure various atmospheric parameters during solar eclipses. The technical specifications of the instruments include:

   - Temperature Sensors: High-precision sensors capable of measuring temperature variations with exceptional accuracy.
   - Density Gauges: Instruments equipped with air pressure sensors and particle counters to measure changes in atmospheric density.
   - Electric and Magnetic Field Detectors: Sophisticated detectors capable of measuring fluctuations in electric and magnetic fields within the ionosphere.
   - Radiation Detectors: Instruments capable of detecting various forms of radiation, including ultraviolet and X-ray radiation, to study the effects of solar radiation on Earth's atmosphere.

These instruments were carefully calibrated and integrated into the payload of the rockets to ensure accurate and reliable data collection during flight.

In conclusion, the APEP project represents a monumental achievement in atmospheric science and space exploration. Through collaborative efforts and innovative research methodologies, the project has advanced our understanding of solar eclipse-induced atmospheric perturbations and paved the way for future advancements in climate science, space weather forecasting, and communication technology. The detailed technical specifications of the rockets and instruments used in the project serve as a testament to the dedication and expertise of the scientists and researchers involved in this groundbreaking endeavor.

www.ingramcontent.com/pod-product-compliance
Lightning Source LLC
Chambersburg PA
CBHW070412230526
45471CB00006B/2768